ENERGY SECTOR STANDARD
OF THE PEOPLE'S REPUBLIC OF CHINA
中华人民共和国能源行业标准

Technical Specification for Acceptance of Environmental Protection in Initial Impoundment for Hydropower Projects

水电工程蓄水环境保护验收技术规程

NB/T 10130-2019

Chief Development Department: China Renewable Energy Engineering Institute

Approval Department: National Energy Administration of the People' s Republic of China

Implementation Date: October 1, 2019

China Water & Power Press

中国水利水电出版社

Beijing 2024

图书在版编目（CIP）数据
水电工程蓄水环境保护验收技术规程 ： NB/T 10130-2019 = Technical Specification for Acceptance of Environmental Protection in Initial Impoundment for Hydropower Projects （NB/T 10130-2019） ： 英文 / 国家能源局发布. -- 北京 ： 中国水利水电出版社, 2024. 8. -- ISBN 978-7-5226-2707-6
Ⅰ. TV512-65
中国国家版本馆CIP数据核字第2024HC3958号

ENERGY SECTOR STANDARD
OF THE PEOPLE'S REPUBLIC OF CHINA
中华人民共和国能源行业标准

Technical Specification for Acceptance of Environmental Protection in Initial Impoundment for Hydropower Projects

水电工程蓄水环境保护验收技术规程

NB/T 10130-2019

（英文版）

Issued by National Energy Administration of the People's Republic of China
国家能源局　发布
Translation organized by China Renewable Energy Engineering Institute
水电水利规划设计总院　组织翻译
Published by China Water & Power Press
中国水利水电出版社　出版发行
Tel: (+ 86 10) 68545888　68545874
sales@mwr.gov.cn
Account name: China Water & Power Press
Address: No.1, Yuyuantan Nanlu, Haidian District, Beijing 100038, China
http: //www.waterpub.com.cn
中国水利水电出版社微机排版中心　排版
北京中献拓方科技发展有限公司　印刷
184mm×260mm　16开本　3印张　95千字
2024年8月第1版　2024年8月第1次印刷
Price（定价）: ¥ 495.00

Introduction

This English version is one of China's energy sector standard series in English. Its translation was organized by China Renewable Energy Engineering Institute authorized by National Energy Administration of the People's Republic of China in compliance with relevant procedures and stipulations. This English version was issued by National Energy Administration of the People's Republic of China in Announcement [2023] No. 4 dated May 26, 2023.

This version was translated from the Chinese Standard NB/T 10130-2019, *Technical Specification for Acceptance of Environmental Protection in Initial Impoundment for Hydropower Projects*, published by China Water & Power Press. The copyright is reserved by National Energy Administration of the People's Republic of China. In the event of any discrepancy in the implementation, the Chinese version shall prevail.

Many thanks go to the staff from the relevant standard development organizations and those who have provided generous assistance in the translation and review process.

For further improvement of the English version, any comments and suggestions are welcome and should be addressed to:

China Renewable Energy Engineering Institute
No. 2 Beixiaojie, Liupukang, Xicheng District, Beijing 100120, China
Website: www.creei.cn

Translating organizations:

POWERCHINA Beijing Engineering Corporation Limited

China Renewable Energy Engineering Institute

Translating staff:

GAO Yan   QI Wen   GUO Jie   LU Bo

YUAN Yuan   JIN Yi   XU Qiulan

Review panel members:

QIE Chunsheng	Senior English Translator
YU Weiqi	China Renewable Energy Engineering Institute
LI Zhongjie	POWERCHINA Northwest Engineering Corporation Limited
LI Kejia	POWERCHINA Northwest Engineering Corporation Limited

CHEN Lei	POWERCHINA Zhongnan Engineering Corporation Limited
DONG Haoping	POWERCHINA Huadong Engineering Corporation Limited
PENG Qidong	China Institute of Water Resources and Hydropower Research
ZHANG Ming	Tsinghua University

National Energy Administration of the People's Republic of China

翻译出版说明

本译本为国家能源局委托水电水利规划设计总院按照有关程序和规定，统一组织翻译的能源行业标准英文版系列译本之一。2023 年 5 月 26 日，国家能源局以 2023 年第 4 号公告予以公布。

本译本是根据中国水利水电出版社出版的《水电工程蓄水环境保护验收技术规程》NB/T 10130—2019 翻译的，著作权归国家能源局所有。在使用过程中，如出现异议，以中文版为准。

本译本在翻译和审核过程中，本标准编制单位及编制组有关成员给予了积极协助。

为不断提高本译本的质量，欢迎使用者提出意见和建议，并反馈给水电水利规划设计总院。

地址：北京市西城区六铺炕北小街 2 号
邮编：100120
网址：www.creei.cn

本译本翻译单位：中国电建集团北京勘测设计研究院有限公司
水电水利规划设计总院

本译本翻译人员：高　燕　齐　文　郭　洁　陆　波
袁　嫄　金　弈　徐秋兰

本译本审核人员：

郄春生　英语高级翻译
喻卫奇　水电水利规划设计总院
李仲杰　中国电建集团西北勘测设计研究院有限公司
李可佳　中国电建集团西北勘测设计研究院有限公司
陈　蕾　中国电建集团中南勘测设计研究院有限公司
董浩平　中国电建集团华东勘测设计研究院有限公司
彭期冬　中国水利水电科学研究院
张　明　清华大学

国家能源局

Announcement of National Energy Administration of the People's Republic of China [2019] No. 4

National Energy Administration of the People's Republic of China has approved and issued 297 sector standards such as *Code for Electrical Design of Photovoltaic Power Projects*, including 105 energy standards (NB), 168 electric power standards (DL), and 24 petrochemical standards (NB/SH).

Attachment: Directory of Sector Standards

National Energy Administration of the People's Republic of China

June 4, 2019

Attachment:

Directory of Sector Standards

Serial number	**Standard No.**	**Title**	**Replaced standard No.**	**Adopted international standard No.**	**Approval date**	**Implementation date**
…						
3	NB/T 10130-2019	Technical Specification for Acceptance of Environmental Protection in Initial Impoundment for Hydropower Projects			2019-06-04	2019-10-01
…						

Foreword

According to the requirements of Document GNKJ [2014] No. 298 issued by National Energy Administration of the People's Republic of China, "Notice on Releasing the Development and Revision Plan of the First Batch of Energy Sector Standards in 2014", and after extensive investigation and research, summarization of practical experience, and wide solicitation of opinions, the drafting group has prepared this specification.

The main technical contents of this specification include: basic requirements, acceptance preparation, acceptance investigation, and site acceptance.

National Energy Administration of the People's Republic of China is in charge of the administration of this specification. China Renewable Energy Engineering Institute has proposed this specification and is responsible for its routine management. Energy Sector Standardization Technical Committee on Hydropower Planning, Resettlement and Environmental Protection is responsible for the explanation of the specific technical content. Comments and suggestions in the implementation of this specification should be addressed to:

China Renewable Energy Engineering Institute
No. 2 Beixiaojie, Liupukang, Xicheng District, Beijing 100120, China

Chief development organizations:

China Renewable Energy Engineering Institute

POWERCHINA Beijing Engineering Corporation Limited

Participating development organization:

POWERCHINA Zhongnan Engineering Corporation Limited

Chief drafting staff:

ZHONG Zhiguo	LI Min	JIN Yi	LIU Guihua
TAN Qilin	LU Bo	CUI Ru	DONG Leihua
WANG Longgao	LI Qianqian	ZHANG Bo	LIU Fei
PAN Li	WANG Mengying	ZHANG Zhiguang	LI Yifei
KONG Yong	CUI Xiaohong	CHANG Yi	ZHANG Shalong
LI Xiang	ZHAO Kun	YAN Jianbo	YU Weiqi

Review panel members:

WAN Wengong	CHEN Yuying	YANG Hongbin	CHEN Guozhu

CHEN Yongbai	YANG Hongwei	ZENG Deyong	XU Yong
CHEN Bangfu	DAI Xiangrong	SHI Jiayue	KOU Xiaomei
JIANG Yueliang	TANG Zhongbo	ZHANG Rong	YIN Xianjun
LI Shisheng			

Contents

1 General Provisions

1.0.1 This specification is formulated with a view to standardizing the work scope, procedures and methods and unifying the technical requirements for the acceptance of environmental protection in initial impoundment for hydropower projects.

1.0.2 This specification is applicable to the acceptance of environmental protection in initial impoundment for hydropower projects.

1.0.3 The acceptance of environmental protection in initial impoundment for hydropower projects shall be based on relevant environmental protection laws and regulations, technical standards, environmental impact assessment (EIA) documents and approval documents, environmental protection design documents and contracts; adhere to the principles of objectivity, impartiality and being focused; make full use of the acceptance results of environmental protection facilities; and coordinate the relationship with the acceptance of environmental protection for completed hydropower projects.

1.0.4 The acceptance of environmental protection in initial impoundment for hydropower projects shall cover the environmental protection facilities to be completed before the initial impoundment and the implementation of protection measures, and place emphasis on the initial impoundment-related layered water intaking facilities, ecological flow discharge facilities, fish passage facilities, rare and key protected plants and old trees transplanting and hazardous waste disposal in the reservoir inundation area, etc.

1.0.5 The acceptance of environmental protection in initial impoundment for hydropower projects shall, on the basis of the acceptance preparation, conduct the acceptance investigation for environmental protection facilities through field inspection and monitoring verification, perform the site acceptance inspection, organize acceptance meeting, and present acceptance opinions.

1.0.6 In addition to this specification, the acceptance of environmental protection in initial impoundment for hydropower projects shall comply with other current relevant standards of China.

2 Basic Requirements

2.0.1 The acceptance of environmental protection in initial impoundment for hydropower projects shall check the environmental protection policies, environmentally sensitive areas, environmental protection objectives, project schemes and environmental protection facilities; identify the environmental management tasks; check the implementation of the "Three Simultaneities" system, i.e. to design, construct and put into operation the environmental protection measures simultaneously with the main works; verify the operation effect of the environmental protection facilities; and put forward the improvement suggestions on the subsequent environmental management and the environmental protection measure implementation.

2.0.2 For the acceptance of environmental protection in initial impoundment for hydropower projects, the following requirements shall be fulfilled:

1 The hydropower complex construction progress meets the initial impoundment requirements of the project. The works in the reservoir area affected by the initial impoundment has been completed, and the resettlement and the reservoir basin clearance have been completed.

2 The environmental protection facilities of the complex structures have been constructed simultaneously with the structures, the facilities for environmental protection and soil and water conservation affected by the initial impoundment have been completed, and the ecological flow discharging plan is applicable during the initial impoundment.

2.0.3 The acceptance of environmental protection in initial impoundment for hydropower projects shall include acceptance preparation, acceptance investigation and site acceptance. The acceptance procedure of environmental protection in initial impoundment for hydropower projects shall comply with Appendix A of this specification.

2.0.4 The acceptance of environmental protection in initial impoundment for hydropower projects shall meet the following requirements:

1 The archives of the environmental protection review, approval and change procedures are complete, and the technical data and the environmental protection archives are complete.

2 The main environmental protection facilities and measures that must be completed before the initial impoundment as required by the EIA documents and approval documents have been implemented, and their normal operating conditions are available.

3 During the construction, all kinds of pollutants are effectively treated or managed and meet the pollution prevention and control objectives specified in the EIA documents; the environmental quality of the construction area meets the environmental quality protection objectives specified in the EIA documents.

4 The quality of the special works such as low-temperature water impact mitigation facilities, ecological flow discharge facilities and fish passage facilities shall meet the requirements of acceptance of the main works.

5 The environmental complaints such as environmental pollution and ecological damage during the construction have been rectified.

2.0.5 The project owner shall organize relevant organizations to conduct the acceptance of environmental protection in initial impoundment for the hydropower project based on the overall arrangement of impoundment. The responsibilities of relevant participants for environmental protection acceptance in initial impoundment for hydropower projects shall comply with Appendix B of this specification.

2.0.6 The project owner shall organize the preparation of the investigation report for acceptance of environmental protection in initial impoundment for hydropower projects. The contents of investigation report for environmental protection acceptance in initial impoundment for hydropower projects should be in accordance with Appendix C of this specification.

3 Acceptance Preparation

3.1 General Requirements

3.1.1 The acceptance preparation for environmental protection in initial impoundment for hydropower projects shall include acceptance work program, documentation and site preparation.

3.1.2 The project owner shall, in accordance with the requirements of the project EIA documents, environmental protection design documents and their examination and approval documents, develop the acceptance work program for environmental protection in initial impoundment based on the construction progress, and conduct the documentation and site preparation before the acceptance.

3.2 Acceptance Work Program

3.2.1 The project owner shall, according to the overall arrangement of the initial impoundment of the hydropower project, determine the work plan for acceptance of environmental protection in the initial impoundment, and complete the acceptance of environmental protection before the initial impoundment.

3.2.2 The project owner shall be responsible for organizing the preparation of the acceptance work program. The acceptance work program should include acceptance procedures, acceptance organization, acceptance scope, and schedule. The scope and focus of environmental protection acceptance in initial impoundment for hydropower projects shall comply with Appendix D of this specification.

3.3 Documentation

3.3.1 The project owner shall organize all relevant organizations to summarize the work before the initial impoundment, and present summary documents.

3.3.2 The project owner shall organize the documents sorting and filing for the acceptance of environmental protection in initial impoundment for hydropower projects. The list of documents for environmental protection acceptance in initial impoundment for hydropower projects should be in accordance with Appendix E of this specification.

3.4 Site Preparation

3.4.1 The project owner shall complete the environmental protection facilities in accordance with the environmental impact statement and its approval document, accomplish the acceptance of the sections or special items of the environmental protection facilities according to the construction progress, and conduct self-check and rectification for environmental protection facilities and environmental protection measures before the acceptance investigation.

3.4.2 The project owner shall conduct rectification for environmental protection facilities and measures according to the comments of acceptance investigation organization and the technical review.

3.4.3 The site preparation shall meet the following requirements:

1 Site access and communication are unobstructed.

2 Environmental protection facilities are in normal operation.

3 Operation and management staff of environmental protection facilities are in place.

4 Acceptance Investigation

4.1 General Requirements

4.1.1 The range of acceptance investigation shall be determined according to the actually affected areas due to the project construction before impoundment and the possibly affected areas due to impoundment taking into account the distribution of environmentally sensitive objects, with emphasis on the keyworks construction area, reservoir inundation area, and possibly affected downstream area.

4.1.2 The acceptance investigation of the environmental protection in initial impoundment shall include:

1 Project construction and engineering design changes.

2 Changes in the environmental protection requirements and external environmental factors such as environmentally sensitive objects.

3 Implementation and effects of environmental protection measures.

4 Environmental impacts caused by construction activities before initial impoundment.

5 Implementation of environmental management.

6 Implementation of environmental monitoring.

7 Public opinions.

8 Conclusions and suggestions.

4.1.3 The acceptance criteria for the environmental protection in the initial impoundment shall be in accordance with the environmental protection standards adopted and the environmental protection requirements set forth in the EIA documents. If, after the approval of the EIA documents, the environmental criteria or environmental functions change, the conformity of the design criteria and the effectiveness of the environmental protection facilities shall be reviewed.

4.1.4 If the acceptance investigation finds that the environmental protection measures are not put in place or the changes have not gone through the formalities as required, the acceptance investigation organization shall timely notify the project owner in writing and put forward suggestions on the implementation of the change formalities and improvement.

4.2 Project Investigation

4.2.1 The project investigation shall cover the river basin where the project is located, the project and the analysis of the environmental reasonableness of the engineering design changes.

4.2.2 The investigation of the river basin where the project is located shall cover:

1 The natural conditions of the river basin, including geological and geomorphological characteristics, hydrographic features, runoff characteristics, river channel morphology, etc.

2 The hydropower planning of the river basin and current situation of cascade development.

4.2.3 The project investigation shall cover:

1 Project geographic location; works composition, project scale and characteristics; keyworks layout and main structures; resettlement implementation; engineering design changes; and accomplishment of project total investment and environmental protection investment.

2 Construction general layout; the location, scale and range of spoil areas, quarries and borrow areas; project construction status; construction method; and the quantities of main works.

3 Project approval process and construction progress, the integrity of approval procedure and review of the project design document, major events during the construction such as the official commencement of the project and river closure, the project physical progress and the acceptance completion of the reservoir basin clearance.

4 Project impoundment scheme, power station operation scheme and reservoir regulation scheme.

5 Changes in project development purpose, reservoir characteristic water level, operation mode, installed capacity, keyworks layout, construction planning and resettlement scheme; and the performance of relevant formalities.

4.2.4 The engineering design changes shall be investigated and the environmental reasonableness shall be analyzed. The form of engineering design changes for hydropower projects should be in accordance with Appendix F of this specification.

4.3 Review of Environmental Protection Requirements

4.3.1 The acceptance investigation shall review the main conclusions and review comments of the EIA or environmental impact retrospective assessment of the hydropower planning in the river basin, with a focus on the requirements related to environmental protection of the project.

4.3.2 The acceptance investigation shall review the environmentally sensitive objects according to the construction characteristics and environmental features of the affected area. When the environmentally sensitive objects change, the investigation scope shall be adjusted accordingly. The form of investigation of environmentally sensitive objects for hydropower projects should be in accordance with Appendix G of this specification.

4.3.3 The acceptance investigation shall review the changes in environmental function zoning, environment criteria and environmental protection objectives.

4.4 Investigation of Environmental Protection Measures

4.4.1 The investigation of environmental protection measures in the initial impoundment shall include:

1 Wastewater and sewage treatment measures during construction.

2 Ecological flow discharge measures.

3 Layered water intaking measures.

4 Fish passage measures.

5 Fish restocking measures.

6 Aquatic habitat protection measures.

7 Rare and key protected plants and animals and old trees protection measures.

8 Reservoir basin clearance and hazardous waste disposal measures.

4.4.2 The investigation of environmental protection measures shall review the location, process and effect of the measures using existing written materials and videos, with emphasis on the implementation of the "Three Simultaneities" system.

4.4.3 The investigation organization shall check and verify the procedural compliance, design conformity, reasonableness of changes, and effectiveness of measures and data integrity of environmental protection measures item by item, and record the results. The checklist of environmental protection facilities in initial impoundment for hydropower projects should be in accordance with the Appendix H of this specification.

4.4.4 The investigation of wastewater and sewage treatment measures in the construction area shall meet the following requirements:

1 Investigate the design conformity of wastewater and sewage treatment facilities, with emphasis on the construction site, process and scale.

2 Investigate the conformity of the construction and operation of wastewater and sewage treatment facilities and the "Three Simultaneities" system.

3 Investigate the amount of wastewater and sewage generated and the amount of wastewater and sewage disposed, the compliance of effluent quality, the discharge path and the treatment effect.

4 Investigate the clearing and disposal of sludge and floating oil generated in the process of wastewater and sewage treatment, including the treatment process, clearing frequency, treatment volume and the destination after treatment.

5 Investigate the professional staffing, assignment of duties, management system and operation records, and analyze the operation and maintenance of wastewater and sewage treatment facilities in the construction area.

6 Investigate the investment in wastewater and sewage treatment facilities.

7 Investigate the complaints about water pollution caused by construction and the improvement of prevention and treatment measures by the participating organizations.

8 Put forward the investigation conclusions and improvement suggestions on wastewater and sewage treatment measures in the construction area.

4.4.5 The investigation of ecological flow discharge measures shall meet the following requirements:

1 Investigate the design conformity of ecological flow discharge measures, with emphasis on the type, layout pattern, location, scale and corresponding scheduling scheme.

2 Investigate and analyze whether the construction and operation of the ecological flow discharge facilities can keep pace with the construction of the main works, the impoundment of the reservoir and the trial operation of the power station.

3 Investigate the construction of ecological flow online monitoring facilities, and analyze if a sound operation and maintenance system for ecological flow discharge measures is established.

4 According to the environmental protection design documents, supervision documents, acceptance documents and operation scheme of the ecological flow discharge measures, give a clear conclusion whether or not the acceptance requirements are satisfied, and put forward the operation requirements of the ecological flow discharge measures for the next stage.

4.4.6 The investigation of layered water intaking measures shall meet the following requirements:

1 Investigate the design conformity of layered water intaking measures, with emphasis on the type, layout, scale, and structure and corresponding operation scheme.

2 Investigate and analyze whether the construction and operation of the layered water intaking measures can keep pace with the construction of the main works, the impoundment of the reservoir and the trial operation of the power station.

3 Investigate the professional staffing, assignment of duties, and the management system, and analyze if a sound operation and maintenance system for layered water intaking measures is established .

4 According to the design documents, supervision documents, acceptance documents and operation scheme of layered water intaking measures, give a clear conclusion whether or not the acceptance requirements are satisfied, and put forward the operation requirements of the layered water intaking measures for the next stage.

4.4.7 The investigation of fish passage measures shall meet the following requirements:

1 Investigate the design conformity of the fish passage facilities, with emphasis on the layout, type, scale and technologies of the fish passage facilities.

2 Investigated and analyze whether the construction and operation of the fish passage measures can keep pace with the construction of the main works, the impoundment of the reservoir and the trial operation of the power station.

3 Investigate the operation plan and review documents of the fish passage facilities, and the professional staffing, assignment of duties and the management system, and analyze if a sound operation and maintenance management system for the fish passage facilities is established.

4 According to the design documents, supervision documents, acceptance documents and operation scheme of fish passage facilities, give a clear conclusion whether or not the acceptance requirements are satisfied, and put forward the operation and effect monitoring requirements of the fish passage facilities for the next stage.

4.4.8 The investigation of fish restocking measures shall meet the following requirements:

1 Investigate the design conformity of fish restocking facilities, with emphasis on the layout, objects, scale and technologies.

2 Investigate the relationship between the construction and operation time of the fish restocking facilities and the construction milestones of the main works, such as river closure and initial impoundment, and the compliance with the schedule in the EIA documents and environmental protection design documents.

3 Investigate the professional staffing, assignment of duties, management system, and water supply guarantee measures, and analyze if a sound operation and maintenance system of fish restocking facilities is established.

4 Investigate the parent fish collection plan, breeding plan, releasing plan and preliminary implementation of fish restocking.

5 Investigate the conformity of the object, amount, size, releasing time and place of fish restocking with the requirements of EIA documents and environmental protection design documents.

6 According to the design, supervision and acceptance documents and the implementation of fish restocking measures, analyze the compliance with the acceptance requirements.

4.4.9 The investigation of aquatic habitat protection measures shall meet the following requirements:

1 Investigate the design conformity of aquatic habitat protection measures, with emphasis on the implementation location, scope, scale, function, and facilities.

2 Investigate the conformity of progress of aquatic habitat protection measures with the "Three Simultaneities" required by milestones such as impoundment and trial operation of the power station.

3 According to the design, supervision and acceptance documents

and operation and maintenance scheme of aquatic habitat protection measures, analyze the compliance with the acceptance requirements.

4.4.10 The investigation of protection measures for rare and key protected animals and plants and old trees shall meet the following requirements:

1 Investigate the design conformity of protection measures for rare and key protected animals and plants and old trees, with emphasis on the objects of protection, standard, location, scale and layout pattern, etc.

2 Investigate the implementation progress of protection measures for rare and key protected animals and plants and old trees, and the relationship with the construction milestones of the main works, such as river closure and initial impoundment, and analyze the rationality of the implementation progress according to the ecological characteristics of rare and key protected animals and plants and old trees.

3 Investigate the professional staffing, assignment of duties, management system, and operation or conservation records of protection measures of rare and key protected animals and plants and old trees, and analyze if a sound operation and maintenance system is established.

4 According to the design, supervision and acceptance documents and operation and maintenance scheme of the protection measures of rare and key protected animals and plants and old trees, and analyze the compliance with the acceptance requirements.

4.4.11 The investigation of reservoir basin clearance and hazardous waste disposal measures shall meet the following requirements:

1 Investigate the design conformity of reservoir basin clearance and hazardous waste disposal measures, with emphasis on the completion of relocation of polluting enterprises and the hazardous waste disposal measures.

2 Investigate the implementation and effect of reservoir basin clearance and hazardous waste disposal measures.

3 According to the design, monitoring and acceptance documents of reservoir basin clearance and hazardous waste disposal measures, give a clear conclusion whether or not the acceptance requirements are satisfied, and put forward suggestions for improvement.

4.5 Investigation of Environmental Impact

4.5.1 The investigation of environmental impact shall analyze the impacts of construction on environment according to the environmental monitoring and supervision data.

4.5.2 The environmental impact investigation shall analyze the difference in environmental impact with the engineering design change.

4.6 Investigation of Environmental Management

4.6.1 The investigation of environmental management shall cover environmental management measures and environmental supervision.

4.6.2 The investigation of environmental management measures shall cover the following:

1 Establishment of environmental protection organizations of the project owner and contractors.

2 Organizational management system and assignment of duties.

3 Formulation and implementation of environmental management regulations.

4 Record of design review, acceptance and launching of environmental protection works.

5 Environmental protection training and publicity during construction.

6 Environmental protection supervision and inspection by the competent authorities.

4.6.3 The investigation of environmental supervision shall cover the following:

1 Establishment of environmental supervision organization for project construction.

2 Organizational management system and assignment of duties of environmental supervision organization.

3 Formulation and implementation of environmental supervision regulations.

4 Scope and tasks of environmental supervision, and supervision requirements for various bid lots of environmental protection works.

5 Investigation of the effectiveness of environmental supervision.

6 Record of events in the environmental supervision process.

4.6.4 The investigation of environmental management measures shall analyze the compliance of environmental management organization, environmental management regulations, and training and publicity in the construction period with the requirements of EIA documents. The effectiveness and deficiency of environmental management during construction shall be evaluated.

4.6.5 The environmental supervision investigation shall analyze the compliance of supervision organization, regulations and work process in the construction period with the requirements of EIA documents. The effect of environmental supervision on the implementation of environmental protection measures shall be evaluated.

4.7 Investigation of Environmental Monitoring

4.7.1 The investigation of environmental monitoring shall check the implementation of environmental monitoring and ecological investigation during construction.

4.7.2 The investigation of environmental monitoring shall cover the following:

1 Environmental monitoring regulations.

2 Environmental monitoring organizations and implementation units.

3 Environmental monitoring scheme.

4 Environmental monitoring facilities and equipment.

5 Environmental monitoring results.

6 Environmental monitoring information management.

4.7.3 The investigation of environmental monitoring shall analyze the reliability of the project environmental monitoring results and the conformity of the monitoring scheme with the EIA documents. When the project scheme or environmentally sensitive objects change, the suggestions on the optimization of environmental monitoring measures shall be put forward.

4.8 Public Opinions Survey

4.8.1 The acceptance investigation organization may conduct the survey and evaluation of public opinions as required.

4.8.2 For the public opinions survey, opinions from local departments of environmental protection, water conservancy, forestry, and land and fishery management should be solicited. When there are many environmentally sensitive protection objects in project construction, inundation of reservoirs and downstream affected areas, direct public consultation should be made by panel discussion.

4.8.3 The respondents for public opinions survey shall include stakeholders directly and indirectly affected by the project construction and the impoundment of the reservoir.

4.8.4 The public opinions survey shall be aware of the public complaints about environmental pollution or disturbance during construction, as well as the corresponding investigation and response, solutions and effects, etc.

4.9 Conclusions and Suggestions

4.9.1 The acceptance investigation shall analyze and summarize the environmental protection work, judge whether the project meets the acceptance conditions for environmental protection in the initial impoundment, and draw the general conclusion.

4.9.2 Suggestions for improvement shall be put forward for the items that fail to meet the requirements of environmental protection in the initial impoundment.

4.9.3 The acceptance investigation shall propose suggestions on environmental protection after initial impoundment.

4.10 Technical Review

4.10.1 The project owner should organize the technical review of the investigation report on environmental protection acceptance in initial impoundment.

4.10.2 The technical review meeting may involve the project owner, designers, construction contractors, EIA document preparation organization, acceptance investigation organization, environmental supervision organization and technical experts.

4.10.3 The technical review opinions should include the following:

1 Main conclusions of the acceptance investigation report.

2 Quality evaluation of the acceptance investigation report.

3 Clear conclusions on whether acceptance criteria are met.

4 Revisions to the acceptance investigation report.

5 Comments and suggestions for improvement.

5 Site Acceptance

5.0.1 The acceptance work of environmental protection in initial impoundment for hydropower projects shall include field inspection, data verification and acceptance meeting.

5.0.2 For the site acceptance, an acceptance team shall be set up, which may be composed of representatives of the project owner, designers, construction contractors, supervision organization, monitoring organizations, EIA document preparation organization, and acceptance investigation organization as well as technical experts. The site acceptance meeting should be planned together with the impoundment acceptance of the project with respect to the participating organizations and personnel. In addition, the design, implementation and operation organizations of special environmental protection measures and other stakeholders may also attend the meeting.

5.0.3 The project owner shall prepare a field inspection plan in advance for the impoundment acceptance, and emphasis of field inspection and verification shall be placed on important sensitive areas and important environmental protection facilities.

5.0.4 The project owner shall organize and coordinate the relevant organizations to prepare and archive the data required for site acceptance according to the requirements, and provide the data to the acceptance team for verification. The data required for site acceptance shall include the following:

1 Acceptance investigation report.

2 Environmental impact statement and approval documents.

3 Summary report of environmental supervision.

4 Summary report on environmental protection by the project owner.

5 Summary reports on environmental protection by the construction contractors.

6 Summary reports on environmental monitoring and ecological investigation.

7 Audio and video data.

5.0.5 The field inspection and acceptance meeting of environmental protection in the initial impoundment shall be organized and hosted by the project owner.

5.0.6 The agenda of the site acceptance meeting should be as follows:

1 Announce the meeting agenda and the members of the acceptance team.

2 Watch the project video.

3 Listen to the work report of the project owner.

4 Listen to the work report of the environmental supervision organization.

5 Listen to the work report of the environmental monitoring organizations.

6 Listen to the work report of the acceptance investigation organization.

7 The acceptance team makes inquiry and discussion, and presents the acceptance opinions document signed by the members of the acceptance team.

8 Announce the acceptance opinions.

5.0.7 The acceptance opinions shall state the conclusion, comments for improvement and subsequent requirements. The pass of environmental protection acceptance shall be conditional on the agreement by at least two-thirds of members of the acceptance team, and the team members shall sign the acceptance opinions document. Opinions on environmental protection acceptance in initial impoundment for hydropower projects should be in accordance with Appendix J of this specification.

Appendix A Acceptance Procedure of Environmental Protection in Initial Impoundment for Hydropower Projects

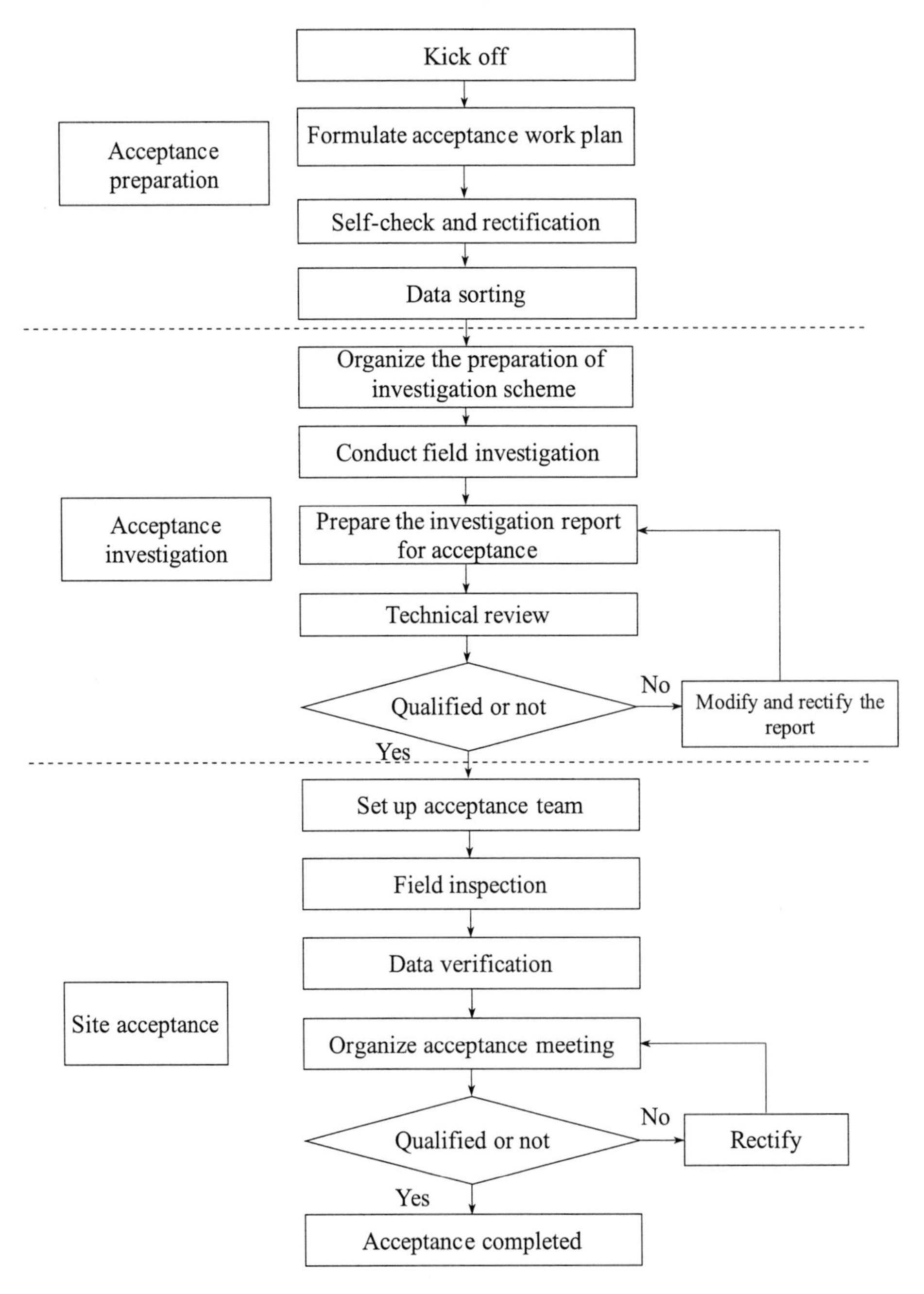

Figure A Acceptance procedure of environmental protection in initial impoundment for hydropower projects

Appendix B Responsibilities of Relevant Participants for Environmental Protection Acceptance in Initial Impoundment for Hydropower Projects

Table B Responsibilities of relevant participants for environmental protection acceptance in initial impoundment for hydropower projects

Organizations	Work stage		
	Acceptance preparation	Acceptance investigation	Site acceptance
Project owner	1) Organize and coordinate involved organizations to sort and file the related data; 2) Prepare acceptance work scheme; 3) Conduct self-check of environmental protection facilities to identify whether the basic conditions for acceptance are met; 4) Organize the environmental protection investigation for acceptance and conduct the preliminary work for acceptance	1) Provide the basic data required by the investigation organization; 2) Make rectification according to the environmental protection rectification opinions from investigation organization; 3) Entrust an organization to hold the technical review; 4) Assist in the technical review meeting; 5) Make rectification according to the environmental protection rectification comments from the technical review organization; report and record the environmental protection facilities that have been adjusted or changed in time	Prepare the field inspection plan in advance, host the field inspection and acceptance meeting

Table B *(continued)*

Organizations	Work stage		
	Acceptance preparation	Acceptance investigation	Site acceptance
Investigation organization for acceptance	–	1) Conduct the preliminary investigation and notify the project owner of the rectification requirements of environmental protection facilities or supplementary opinions; 2) According to the relevant laws and regulations and technical requirements, prepare the acceptance investigation report, and put forward the acceptance conclusions and opinions; 3) Revise and improve the acceptance investigation report according to the review opinions	Prepare the data, and assist in field inspection and acceptance
Technical review organization	–	Organize and host the technical review meeting, conduct the technical review of the investigation results, and put forward the review opinions	–
Designer	Assist the project owner to prepare data and to conduct self-check and preliminary inspection of environmental protection facilities	1) Cooperate with the investigation organization and provide relevant design data; 2) Cooperate with the technical review organization	Cooperate with field inspection and acceptance

Table B *(continued)*

Organizations	Work stage		
	Acceptance preparation	**Acceptance investigation**	**Site acceptance**
EIA document preparation organization	–	–	Cooperate with field inspection and acceptance
Environmental supervision organization	Assist the project owner to prepare data and to conduct self-check and preliminary inspection of environmental protection facilities	1) Cooperate with the investigation organization, provide the environmental supervision data, and prepare the environmental supervision summary report; 2) Cooperate with the technical review organization	Cooperate with field inspection and acceptance
Environmental monitoring organizations	–	–	Cooperate with field inspection and acceptance
Construction contractors	Assist the project owner to prepare data and to conduct self-check and preliminary inspection of environmental protection facilities	1) Cooperate with the investigation organization, provide the operation and management data of environmental protection facilities, and prepare the summary report of construction environmental protection; 2) Cooperate with the technical review organization	Cooperate with field inspection and acceptance

Appendix C Contents of Investigation Report for Environmental Protection Acceptance in Initial Impoundment for Hydropower Projects

Foreword

1 Introduction

1.1 Background

1.2 Preparation Basis

1.3 Purpose and Principle

1.4 Scope and Focus

1.5 Methods

1.6 Environmentally Sensitive Objects

1.7 Acceptance Criteria

1.8 Work Process

2 Project Investigation

2.1 Overview of River Basin Planning and Development

2.2 Project Overview

2.3 Project Construction

2.4 Engineering Design Changes

3 Review of Environmental Impact Statement

3.1 Environmental Impact Assessment Process

3.2 General Conclusions of Environmental Impact Statement

3.3 Approval Document of Environmental Impact Statement

4 Investigation of Environmental Protection Measures Implementation

4.1 Overview

4.2 Water Environment Protection

4.3 Aquatic Ecosystem Protection

4.4 Terrestrial Ecosystem Protection

4.5 Solid Waste Treatment and Disposal

4.6 Atmospheric Environment Protection

4.7 Acoustic Environment Protection

4.8 Environmental Protection for Resettlement

4.9 Investment in Environmental Protection

4.10 Summary

5 Environmental Impact Investigation

5.1 Overview

5.2 Water Environment Impact

5.3 Aquatic Ecosystem Impact

5.4 Terrestrial Ecosystem Impact

5.5 Soil Environment Impact

5.6 Atmospheric Environment Impact

5.7 Acoustic Environment Impact

5.8 Environmental Impact of Resettlement

5.9 Summary

6 Investigation of Environmental Risk Management Measures

6.1 Environmental Risk Factors

6.2 Environmental Risk Accidents and Impact

6.3 Environmental Risk Prevention Measures

6.4 Emergency Preparedness Plan for Environmental Risks

7 Investigation of Environmental Management and Environmental Monitoring

7.1 Environmental Management Measures

7.2 Environmental Supervision

7.3 Environmental Monitoring

8 Public Opinion Survey

8.1 Investigation Purposes and Scope

8.2 Investigation Objects and Methods

8.3 Investigation Result Analysis

8.4 Investigation Conclusions

9 Conclusions and Suggestions

9.1 Conclusions

9.2 Suggestions

Attachments:

1 Approval Documents of Environmental Impact Assessment Documents

2 Review Documents of Environmental Protection Design Documents

Attached Figures:

1 Project Location Map

2 River Basin Hydrographic Map

3 Environmental Protection Sensitive Objects Distribution Map

4 Keyworks Layout

5 Construction General Layout

6 General Layout of Environmental Protection Measures

7 Layout of Environmental Protection Facilities

Appendix D Scope and Focus of Environmental Protection Acceptance in Initial Impoundment for Hydropower Projects

Table D Scope and focus of environmental protection acceptance in initial impoundment for hydropower projects

S/N	Scope		Focus
1	Project and engineering design changes	1) Project implementation; 2) Engineering design changes and the reasons; 3) Environmental reasonableness analysis of engineering design changes	Engineering design changes and reasons analysis
2	Performance of formalities	1) Project approval documents; 2) Implementation of project EIA regulatory system; 3) Performance of engineering design change formalities	Legitimacy and compliance of project construction procedures
3	Implementation and operation of environmental protection measures	1) Surface water environment protection measures; 2) Aquatic ecosystem protection measures;	1) Construction of temporary discharging facilities for initial impoundment, permanent discharging facilities for ecological flow, and automatic measurement and reporting, automatic transmission and storage systems for discharging ecological flow;

Table D *(continued)*

S/N	Scope		Focus
3	Implementation and operation of environmental protection measures	3) Terrestrial ecosystem protection measures; 4) Acoustic environment protection measures; 5) Air environment protection measures; 6) Solid waste treatment and disposal measures; 7) Other environmental protection measures	2) Supply of water for downstream sensitive protection objects by the impoundment process and the operation and regulation mode of the reservoir; 3) For reservoirs involving low temperature water, attention shall be paid to the construction of layered water intaking facilities; 4) Construction of fish passage facilities; 5) Construction of fish restocking station and its operation; 6) Implementation of habitat protection measures; 7) Protection and transplanting of rare and key protected plants and old trees; 8) Implementation of protection and management measures, such as migration route and artificial alternative habitats for rare and key protected animals; 9) Implementation of the removal of polluting enterprises and the disposal of hazardous waste involved in the reservoir basin clearance
4	Environmental management system	1) Implementation of the environmental management measures; 2) Implementation of the environmental supervision; 3) Implementation of environmental monitoring; 4) Risk accident prevention and emergency response measures	–

Table D *(continued)*

S/N	Scope		Focus
5	Public participation	1) Complaints and their handling concerning public environmental protection during construction; 2) Survey conclusions of public opinions	–
6	Subsequent environmental protection work plan and suggestion	1) Impoundment scheme and reservoir operation and regulation scheme; 2) Power station operation plan; 3) Subsequent environmental protection work plan	–

Appendix E List of Documents for Environmental Protection Acceptance in Initial Impoundment for Hydropower Projects

Table E List of documents for environmental protection acceptance in initial impoundment for hydropower projects

S/N	Data types	Data
1	EIA documents	River hydropower planning environmental impact statement, river hydropower development environmental impact tracking assessment document, project environmental impact statement
		Special investigation on terrestrial ecology, aquatic ecology and environmental status monitoring during EIA
2	Design documents	Feasibility study report of main works
		Implementation plan of the "Three Simultaneities", overall design report of environmental protection, and special design report of environmental protection facilities
		Construction drawings of environmental protection facilities, and their design changes and construction instructions
		Bidding documents and bids for construction of environmental protection facilities
		As-built drawing, settlement at completion and other data of environmental protection facilities
3	Technical review documents	Approval documents of environmental impact statement
		Review opinions on environmental protection facilities and approval documents for design changes
4	Construction data	Record of events of project construction
		Contracts and agreements for construction of environmental protection facilities
		Acceptance certificate for sections of environmental protection facilities or acceptance documents of special facilities, or self-check and preliminary inspection report
		List of works for acceptance, list of uncompleted works, construction arrangement and completion period of uncompleted works, existing problems and solutions
		Summary report on environmental protection works construction
		Quality evaluation report on environmental protection facilities

Table E *(continued)*

S/N	Data types	Data
5	Monitoring and supervision data	Summary report on environmental protection supervision of main works
		Summary report on environmental protection facilities supervision
		Environmental protection monitoring report
6	Special item acceptance and operation data	Trial operation records, monitoring and investigation data of environmental protection facilities
		Operation summary report
		Acceptance data of soil and water conservation facilities, environmental protection facilities and resettlement
7	Management documents	Acceptance documents of environmental protection facilities and operation specification of environmental protection facilities
		Project environmental management regulations and organization setup documents
		Documents related to project risk prevention measures and emergency preparedness plan and organization setup documents
8	External documents	Written comments of previous inspections from environmental protection authorities and other competent authorities, and rectifications

Appendix F Form of Engineering Design Changes for Hydropower Projects

Table F Form of engineering design changes for hydropower projects

Item	Approved scheme	Optimized scheme	Main adjustments	Reason for adjustment	Major changes or not	Performance of change formalities	Possible environmental impacts caused by scheme adjustment
Main works							
Construction layout							
Environmental protection measures							

Appendix G Form of Investigation of Environmentally Sensitive Objects for Hydropower Projects

Table G Form of investigation of environmentally sensitive objects for hydropower projects

Environmental elements	EIA period			Project implementation period			Environmental protection measures
	Environmentally sensitive objects	Relation with the project	Protection requirements	Environmentally sensitive objects	Relation with the project	Protection requirements	
Water environment							
Terrestrial ecosystem							
Aquatic ecosystem							
…							

Appendix H Checklist of Environmental Protection Facilities in Initial Impoundment for Hydropower Projects

Table H Checklist of environmental protection facilities in initial impoundment for hydropower projects

Item	Environmental impact statement requirements	Approval opinions	Implementation	Verification conclusion				
				Compliance of procedure	Conformity of design	Reasonableness of change	Effectiveness of measures	Integrity of data
Wastewater and sewage treatment facilities in construction area								
1								
2								
…								
Ecological flow discharge facilities								
1								
2								
…								
Low-temperature water impact mitigation facilities								
1								

Table H *(continued)*

Item	Environmental impact statement requirements	Approval opinions	Implementation	Verification conclusion				
				Compliance of procedure	Conformity of design	Reasonableness of change	Effectiveness of measures	Integrity of data
2								
…								
Fish passage facilities								
1								
2								
…								
Fish restocking facilities								
1								
2								
…								
Aquatic habitat protection measures								
1								
2								
…								
Protection facilities for rare animals and plants								

Table H *(continued)*

Item	Environmental impact statement requirements	Approval opinions	Implementation	Verification conclusion				
				Compliance of procedure	Conformity of design	Reasonableness of change	Effectiveness of measures	Integrity of data
1								
2								
…								
Reservoir basin clearance measures								
1								
2								
…								
Hazardous waste disposal measures								
1								
2								
…								
Other environmental protection measures								
1								
2								
…								

Appendix J Opinions on Environmental Protection Acceptance in Initial Impoundment for Hydropower Projects

Table J Opinions on environmental protection acceptance in initial impoundment for hydropower projects

Project name		**Location**	
Acceptance time		**Project owner**	
Acceptance investigation organization		**EIA document preparation organization**	
Acceptance opinions	1 Brief introduction to project construction		
	2 Engineering design changes		
	3 Environmental protection measures implementation and commissioning		
	4 Environmental impacts of project construction		
	5 Acceptance conclusion and requirements		
Signature of acceptance team members			

Explanation of Wording in This Specification

1 Words used for different degrees of strictness are explained as follows in order to mark the differences in executing the requirements of this specification.

1) Words denoting a very strict or mandatory requirement:

"Must" is used for affirmation; "must not" for negation.

2) Words denoting a strict requirement under normal conditions:

"Shall" is used for affirmation; "shall not" for negation.

3) Words denoting a permission of a slight choice or an indication of the most suitable choice when conditions permit:

"Should" is used for affirmation; "should not" for negation.

4) "May" is used to express the option available, sometimes with the conditional permit.

2 "Shall meet the requirements of..." or "shall comply with..." is used in this specification to indicate that it is necessary to comply with the requirements stipulated in other relative standards and codes.